Opportunities for Systems Engineering to Contribute to Durability and Damage Tolerance of Hybrid Structures for Airframes

Jean R. Gebman

Prepared for the United States Air Force

PROJECT AIR FORCE

The research described in this report was sponsored by the United States Air Force under Contract FA7014-06-C-0001. Further information may be obtained from the Strategic Planning Division, Directorate of Plans, Hq USAF.

Library of Congress Cataloging-in-Publication Data

Gebman, J. R.
    Opportunities for systems engineering to contribute to durability and damage tolerance of hybrid structures
for airframes / Jean R. Gebman.
        p. cm.
    Includes bibliographical references.
    ISBN 978-0-8330-4202-6 (pbk. : alk. paper)
    1. Airframes—Design and construction. 2. Structural dynamics. 3. Airframes—Materials. 4. Composite
materials. 5. Airplanes, Military—Design and construction. 6. Fault tolerance (Engineering) 7. Systems
engineering. I. Title.

TL671.6.G43 2007
629.134'31—dc22

2007039674

The RAND Corporation is a nonprofit research organization providing objective analysis and effective solutions that address the challenges facing the public and private sectors around the world. RAND's publications do not necessarily reflect the opinions of its research clients and sponsors.

RAND® is a registered trademark.

Published 2008 by the RAND Corporation
1776 Main Street, P.O. Box 2138, Santa Monica, CA 90407-2138
1200 South Hayes Street, Arlington, VA 22202-5050
4570 Fifth Avenue, Suite 600, Pittsburgh, PA 15213-2665
RAND URL: http://www.rand.org/
To order RAND documents or to obtain additional information, contact
Distribution Services: Telephone: (310) 451-7002;
Fax: (310) 451-6915; Email: order@rand.org

Preface

Concepts of durability and damage tolerance provide the foundation for the Air Force's Aircraft Structural Integrity Program (ASIP), which plays a critical role in ensuring the airworthiness of Air Force aircraft. ASIP provides a common framework for managing and engineering the conceptualization, development, production, operation, and sustainment of airframes. Tailoring and implementing ASIP tasks to best fit the life-cycle needs and circumstances of individual programs are areas in which the methods and practices of systems engineering have much to offer. As the complexity of hybrid airframes increases, the opportunities for systems engineering to add life-cycle value will increase further. Hybrid structures not only incorporate multiple types of materials, but their components often serve multiple functions in addition to transmitting structural loads.

The author prepared this report for the 10th Joint DoD/NASA/FAA Conference on Aging Aircraft. The report draws from his work on aging aircraft, which continues to be sponsored by the U.S. Air Force as a study within RAND Project AIR FORCE. That continuing effort, "Status and Risk Assessments for Aging Aircraft," is sponsored by Lt Gen Donald J. Hoffman, Military Deputy, Office of the Assistant Secretary of the Air Force for Acquisition, Headquarters U.S. Air Force (SAF/AQ); and Lt Gen Raymond E. Johns, Jr., Deputy Chief of Staff for Strategic Plans and Programs, Headquarters U.S. Air Force (AF/A8). The report also draws on the author's contributions to the development of the systems-engineering curriculum for the Air Force Institute of Technology and the University of California at Los Angeles. The report is intended to be of interest to those responsible for tailoring and implementing ASIP tasks. The report was written on the author's personal time and prepared for presentation at the conference with the assistance of Project AIR FORCE.

Previous work on ASIP includes

- Yool Kim, Stephen Sheehy, and Darryl Lenhardt, *A Survey of Aircraft Structural-Life Management Programs in the U.S. Navy, the Canadian Forces, and the U.S. Air Force*, Santa Monica, Calif.: RAND Corporation, MG-370-AF, 2006, 2006.

RAND Project AIR FORCE

RAND Project AIR FORCE (PAF), a division of the RAND Corporation, is the U.S. Air Force's federally funded research and development center for studies and analyses. PAF provides the Air Force with independent analyses of policy alternatives affecting the development, employment, combat readiness, and support of current and future aerospace forces. Research is conducted in four programs: Aerospace Force Development; Manpower, Personnel, and Training; Resource Management; and Strategy and Doctrine. Integrative research projects and work on modeling and simulation are conducted on a PAF-wide basis.

Additional information about PAF is available on our Web site:
http://www.rand.org/paf/

Contents

Figures

Summary

Although a general approach to fielding durable, damage-tolerant structures has been well defined for several decades for metal airframes, the rising use of other materials and the growing role of hybrid structures in airframes are creating a need to tailor the general approach to deal with new damage mechanisms. This has created opportunities for systems engineering to contribute to the tailoring and implementation of the general approach to hybrid structures for airframes. Such implementation can help ensure that an appropriate sequence of investments is made in time to support key decisions related to the research, design, development, test, manufacturing, and sustainment of airframes that have hybrid structures. As industry and operators are tailoring the implementation of the general approach, this may be a good time to pause and consider how well materials engineers, structural engineers, and systems engineers are performing as a team in assuring the durability and damage tolerance of hybrid structures for airframes over their life cycles.

To support such considerations, this report starts by summarizing the Air Force's general approach to developing and sustaining durable, damage-tolerant structures for airframes (see pp. 15–20). Although the details of the approach evolved during an era of metal airframes, its general framework is broadly applicable to airframes in general. Because hybrid structures that have multiple classes of materials are accounting for a growing proportion of the structural assemblies in modern airframes and because they introduce new challenges for durability and damage tolerance, this report explores how systems-engineering efforts may help tailor implementation of the general approach to hybrid structures for airframes.

The report also identifies technical and programmatic considerations that need to be addressed by a systems-engineering approach (see pp. 21–24). Next, the report identifies opportunities for materials engineers and structural engineers to collaborate with systems engineers in ensuring the durability and damage tolerance of hybrid structures in airframes (see pp. 25–27). Finally, it describes a candidate framework for facilitating such collaboration (see pp. 29–34). Such a framework may provide a useful basis for considering and continuously improving the team performance of the materials engineers, structural engineers, and systems engineers who are responsible for ensuring the durability and damage tolerance of hybrid structures over an airframe's life cycle.

Acknowledgments

The process of writing the report included informal reviews and ideas from many colleagues, including Russell Alford, Elliot Axelband, Natalie Crawford, Joseph Gallagher, Edward Keating, Yool Kim, and Richard Kinzie. The author thanks each of these contributors. The author also thanks Robert Ernst and Giles Smith for their formal reviews.

The author further acknowledges the importance of past work on many projects related to the acquisition and sustainment of weapon systems. Associations with many colleagues at RAND over the years contributed to shaping a context within which the present report was written. The author also thanks Michael Neumann and the editor for their assistance in preparing and editing the manuscript.

Abbreviations

AD	airworthiness directive
AF/A8	Deputy Chief of Staff for Strategic Plans and Programs, Headquarters, U.S. Air Force
AFPD	Air Force Policy Directive
ASIP	Aircraft Structural Integrity Program
DoD	Department of Defense
FAA	Federal Aviation Administration
GEITA	Government Electronics and Information Technology Association
IEEE	Institute of Electrical and Electronics Engineers
INCOSE	International Council of Systems Engineers
MIL-STD	military standard
NASA	National Aeronautics and Space Administration
NRC	National Research Council
NTSB	National Transportation Safety Board
PAF	RAND Project AIR FORCE
SAF/AQ	Assistant Secretary of the Air Force for Acquisition, Headquarters, U.S. Air Force
SEMP	Systems Engineering Management Plan
SEP	Systems Engineering Plan
SERP	Systems Engineering Requirements Plan
USAF	U.S. Air Force
USN	U.S. Navy

CHAPTER ONE

Introduction

Systems engineering now has important opportunities to contribute to the durability and damage tolerance of hybrid structures because the effective orchestration of the total engineering effort for such structures increasingly needs the kind of interdisciplinary systems approach that the activities and tools of systems engineering can provide.[1] Contributing to the rising need are

1. a growing reliance on hybrid structures for the primary load paths in airborne vehicles[2]
2. the growing complexity of these structures, especially in military vehicles, in which structural elements are being tasked to perform a variety of nonstructural functions.

Background

Recent trends illustrate how increasing complexity is contributing to a rising need for an interdisciplinary approach to the design, manufacture, and sustainment of modern structures in many airborne vehicles the military uses (Figure 1.1). One existing field, systems engineering, offers tools and activities that can help formulate just such an approach.

Trends

For metal parts and metal structures, the engineering practices for fielding durable, damage-tolerant airframes continue to mature. Metal fatigue and metal corrosion have received much of the emphasis. Nonmetal materials, hybrid materials, and hybrid structures can require different or additional practices, however.

Meanwhile, design of hybrid structures is expanding rapidly. As new durability and damage-tolerance issues are emerging, engineering practices continue to evolve to provide the right mix of practices for ensuring suitable durability and damage tolerance. The right mix is

---

[1] Some departments and agencies of the U.S. federal government use the term *systems engineering* to refer to a set of engineering practices and methods that apply to public-sector projects. The private sector sometimes refers to such practices and methods as *product-development engineering*.

[2] This report uses *hybrid structure* to refer to a structure that (1) carries flight-essential (primary) loads and (2) is fabricated from a mix of different classes of material, of which one class may be metal. The term hybrid material refers to nonhomogeneous material that is formed from multiple types of source material. *Primary load paths* run through structures that carry loads essential to safe flight. *Secondary load paths* run through structures that do not carry loads that are essential to safe flight. Common examples of such secondary structure may include the leading edge of a wing and the fairing between a wing and a fuselage.

Figure 1.1
Recent Trends in Structural Design and a Way Ahead for Overcoming the Challenges

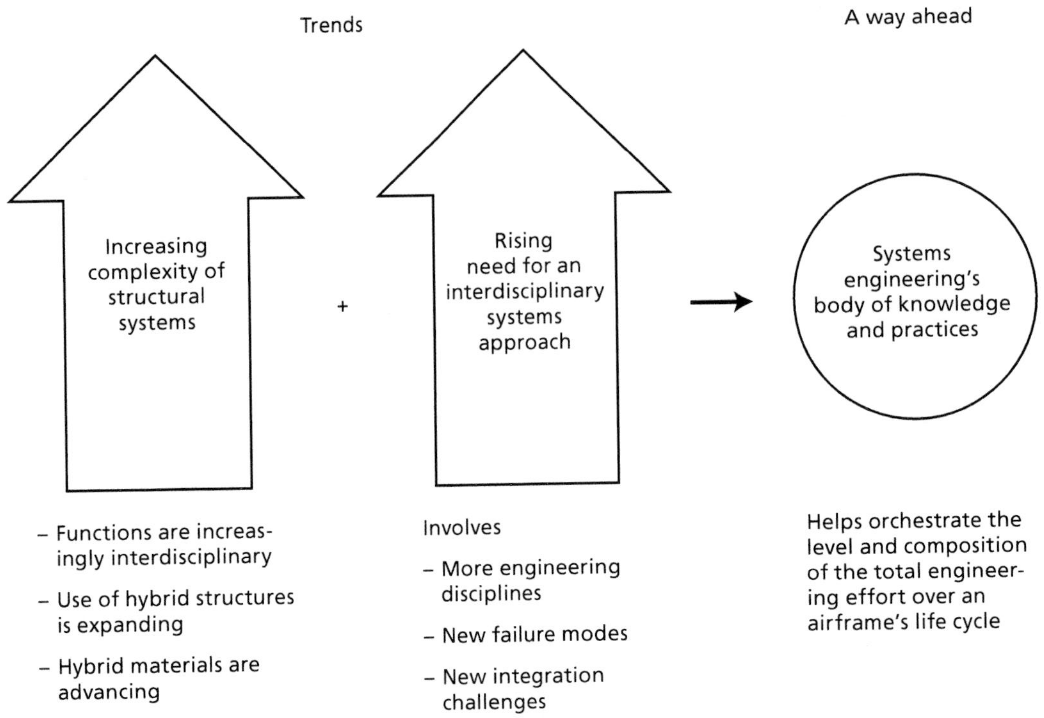

Trends   A way ahead

Increasing
complexity of
structural
systems

+

Rising
need for an
interdisciplinary
systems
approach

Systems
engineering's
body of knowledge
and practices

– Functions are increas-
  ingly interdisciplinary

– Use of hybrid structures
  is expanding

– Hybrid materials are
  advancing

Involves

– More engineering
  disciplines

– New failure modes

– New integration
  challenges

Helps orchestrate the
level and composition
of the total engineer-
ing effort over an
airframe's life cycle

RAND *TR489-1.1*

influenced by material characteristics, material applications, aircraft use, operating environments, and choices about how to balance development and sustainment burdens.

Increasing Complexity. Growing reliance on hybrid structures is increasing complexity for many military systems as structures evolve into multifunction systems with embedded antennas, infrared windows, optical windows, smart skins, and morphing surfaces. This contributes to complexity in several ways:

- Optimizing the design of multifunction components requires an interdisciplinary approach to design that explores trade-offs among different engineering disciplines.
- Materials that are manufactured from multiple types of source materials (hybrid or composite materials) contribute to complexity.
- Structural assemblies that are fabricated from multiple classes of materials (hybrid structures) are another source of complexity, especially where dissimilar materials are joined.

Rising Need for an Interdisciplinary Systems Approach. Growing reliance on hybrid structures is also raising the need for an interdisciplinary systems approach:

- The growing number of engineering disciplines that have an interest in the design of structural assemblies is raising the need to employ formal methods for managing interdisciplinary interactions.
- Introducing additional functions increases integration requirements and challenges.

- Introducing additional functions also introduces additional functional-failure modes that need to be addressed in interdisciplinary processes for design, manufacture, and sustainment.

During the 1970s, the Air Force developed and started applying a standard set of practices for addressing metal fatigue in airframe structures.[3] These practices reflect a systems approach and form an element of the Air Force's systems-engineering approach. Thus far, however, systems-engineering activities and tools have not gone much beyond what is reflected in MIL-STD-1530C. Thus, there is an opportunity to apply such activities and tools further in tailoring the practices defined in the military standard. Such tailoring is now required by Air Force Policy Directive (AFPD) 63-10. As a practical matter, the nature of many hybrid structures is such that tailoring would be required to accommodate peculiar needs that can arise with hybrid structures.

A Way Ahead

The increasing complexity of structural systems, combined with the rising need for an interdisciplinary systems approach, is generating new systems-engineering requirements and opportunities, particularly those that can help orchestrate the level and composition of the total engineering effort over the course of an airframe's life cycle. Even as hybrid structures evolve from their former supporting roles—providing secondary load paths to their emerging leading roles of providing primary load paths, as in the Boeing 787—the case for an expanded role for systems-engineering efforts already has arrived. Furthermore, old definitions of durability and damage tolerance are being overtaken by events.

Today, durability needs to include durability for each of the structure's functions. Similarly, assessments of damage tolerance need to examine the damage tolerance of each function of a structural component. Just as a small crack must not result in the catastrophic loss of load-carrying capacity, a small scratch or crack must not cause unacceptable degradation of a structural component's ability to provide boundary layer control; serve as a heat shield, heat radiator, or antenna; or fulfill some other important function.

As the functionality of structural components expands, so does their complexity and cost. The need to ensure sufficient engineering effort to "get it right" will rise, and the necessary investment in systems engineering will likewise rise. A whole new world is emerging in which engineers will have to think differently about the design, development, and lifetime management of structural systems.

Thus, this report addresses the problem of how to best tailor good practices for developing, fielding, and sustaining hybrid structures that are durable and damage tolerant. The report focuses on a class of opportunities that deal with improving communication and collaboration across three major disciplines of engineering: materials, structures, and systems (Figure 1.2). Each of these major disciplines includes significant fields of specialization that are also within the area of interest for this report. These fields include design engineering, manufacturing engineering, and sustainment engineering. A well-engineered design must be supported by well-engineered manufacturing processes, including quality control. Likewise, a well-produced product must be supported by sound sustainment engineering throughout the product's life cycle.

---

[3] These practices are now described in Military Standard 1530C (MIL-STD-1530C).

Figure 1.2
Scope and Time Horizons of Interest

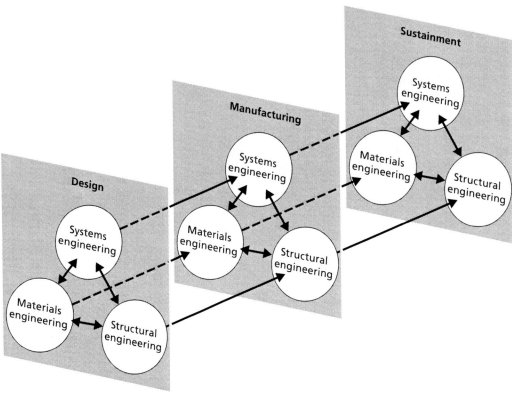

RAND *TR489-1.2*

Interactions among activities occurring at different times also are important. Design choices, for example, may trade a short-term inconvenience during manufacturing for a long-term benefit for durability. Design choices affect durability and damage tolerance both directly and indirectly. A direct effect is the measurable mechanical properties of a selected material. The ability to inspect and repair structural details is one example of an indirect effect; the ability to verify manufacturing and repair quality is another.

A Sampler of Views About Systems Engineering

In terms of actual tasks, the term *systems engineering* takes on a variety of meanings depending on the context. Following is a sampler of views, each of which is applicable to the matters addressed in this report.

A Customer View of Systems Engineering

Recognizing the importance of systems-engineering work to the effective acquisition and sustainment of weapon systems, the Department of Defense (DoD) has a source-selection process that includes evaluation of a contractor's Systems Engineering Management Plan (SEMP) (see U.S. Air Force, 2007, p. 112). Although the DoD does not tell contractors how to conduct their systems-engineering work, it does advise bidders that they can draw guidance from the

DoD's *Systems Engineering Plan (SEP), Preparation Guide* (DoD, 2006). On page 4, that guide states that the

> SEP is the blueprint for the conduct, management, and control of the technical aspects of an acquisition program from conception to disposal, i.e., how the systems engineering process is applied and tailored to meet each acquisition phase objective. The process of planning, developing, and coordinating systems engineering and technical management forces thoughtful consideration, debate, and decisions to produce a sound systems engineering strategy for a program commensurate with the program's technical issues, life cycle phase, and overall objectives.

Although the SEP is the program manager's plan, the government and contractor(s) often develop it jointly. DoD's guidance document states that the SEP should address such matters as the amount of engineering effort, the work products, and the schedule required to achieve the system's requirements. The guidance further indicates the SEP should convey the core information needed to understand the technical approach planned for the program, including

- the technical issues and risks
- who has responsibility and authority for managing the technical issues and risks
- the processes and tools that will be used to address the technical issues and risks
- how the process will be managed and controlled
- how the technical effort will be linked to the overall management of the program.

A Company View of Systems Engineering

Although the systems-engineering work described in this report may be performed by an engineer with a degree in systems engineering or systems architecture, it also may be performed by an engineer who first gained experience in one or more functional-engineering disciplines before receiving company-sponsored training in systems engineering. When a company designs and produces a series of similar products over time, systems engineering may become such an integral part of the company's work processes that it is no longer labeled as such. Some companies, for example, refer to such work as *product-development engineering*.

An Engineering View of Systems Engineering

From the perspective of those who actually conduct systems engineering (whether or not that is what their company calls it), one can think of systems engineering as the systematic decomposition of a system into progressively smaller subsystems, assemblies, and parts, followed by the systematic assembly of pieces, assemblies, and subsystems to produce a functioning system that satisfies the customers' needs over the system's intended period of service.

Both decomposition and assembly must be performed with great care to make the best use of time and resources:

- **Decomposition.** Effectively breaking a system down into pieces that can be developed in parallel by different teams of people requires a combination of technical knowledge of the work to be done, technical knowledge of the capacities of organizations and people, technical understanding of the technical interfaces among teams, and experience. The decomposition process results in the product architecture and the work breakdown structure for the development and manufacturing of the product. The product architecture defines the product's elements at each level of detail, the function of each element, and all the interfaces each element has with other elements.

- **Assembly.** Effective integration of pieces into assemblies, assemblies into subsystems, and subsystems into the final system requires meticulous oversight of the sufficiency of interface control documents to ensure that all pieces, assemblies, and subsystems come together with minimum integration problems. It also requires technical knowledge of the risks at each level of assembly and the ability to make technical judgments about the nature and extent of cost-effective testing at each level of assembly.

One very difficult systems-engineering choice is deciding how much testing is enough at each level of assembly. For example, is it better to invest more time and resources in testing the properties of a new material in the laboratory, or is it better to save that amount of time and resources for full-scale testing in a simulated operational environment? Is it wiser to do some of both, but not as much as either the materials engineer or the structural engineer might prefer? Is there a totally different functional area that presents even greater risks and that, therefore, may be in greater need of scarce resources for testing? Helping assess comparative risks and comparative returns on investment from different engineering activities is a key element of the systems-engineering function.

Engineers in this field have developed a variety of standard practices and tools to assist the systematic gathering of information that can help inform their judgments and choices. Although some companies closely hold such material, because of the competitive advantage it offers, several handbooks are now available that describe many of the more common practices and tools.[4]

Hybrid Structures

By 2002, the Boeing Company had become the world's largest producer and user of nonmetal parts for aerospace applications, spending about $300 million annually for raw material and about $1.7 billion annually for manufacturing (Hahn, 2002a). As industry's use of nonmetal material in aircraft structures continues to expand, both industry and the government strongly seek to ensure that new aircraft have durable, damage-tolerant structures.

For commercial aircraft, the U.S. Federal Aviation Administration (FAA) has established a system of regulations and independent oversight that ensures the continuing airworthiness of aircraft it has certified. The FAA system also ensures that each individual aircraft operated commercially within the United States has a current airworthiness certificate. The FAA system

---

[4] Because failure was not an option, the naval nuclear propulsion program is an example of a program that invested in a very strong set of systems-engineering practices; see Duncan, 1990; Rockwell, 1992; and Rickover, 1979. The ballistic missile programs and the continental air defense program also evidenced a strong application of systems-engineering practices; see Sapolsky, 1972; Beard, 1976; and Baum, 1981. Textbooks on systems engineering at that time focused more on the mathematics than on the strong methods of technical direction that the cited programs employed; see, for example, Porter, 1968; Sage and Melsa, 1971; and Sage, 1977. Contemporary texts focus more on methods of technical architectures, organization of technical efforts, and technical direction; see, for example, Blanchard, 1998; Buede, 2000; and Maier, 2000. Handbooks for systems-engineering practices have been developed by the DoD (Defense Systems Management College, 2001); the U.S. Air Force (Air Force Space and Missile Systems Center, 2004); the National Aeronautics and Space Administration, 1995; the Institute of Electrical and Electronics Engineers, 1998; the Government Electronics and Information Technology Association, 2003; and the International Council of Systems Engineers, 2000.

thereby provides a significant incentive for the design, manufacture, and sustainment of durable, damage-tolerant aircraft.[5]

Organization of This Report

Because it faces different needs and circumstances, DoD takes different approaches to ensure the durability and damage tolerance of military aircraft.[6] This report first addresses the standard practices that comprise the Air Force's general approach to fielding durable, damage-tolerant structures (Chapter Two). It then addresses the systems-engineering activities and tools that can be used to tailor the standard practices to the specific circumstances of individual fleets of airborne vehicles (Chapter Three). Effective application of these activities and tools, however, also requires overcoming technical and programmatic challenges. Chapter Four describes these challenges, and Chapter Five addresses opportunities for overcoming them. Chapter Six describes a framework that could provide a way ahead for pursuing the identified opportunities with a coherent plan of action. Chapter Seven concludes with our key findings and recommendations. Figure 1.3 outlines the main elements of these chapters.

---

[5] The U.S. government has designated the FAA as an airworthiness authority with responsibilities that include setting and enforcing standards for aircraft and air carriers providing commercial services; tracking where airworthiness problems have occurred in aircraft registered in the United States; and issuing airworthiness directives (ADs) when (1) an unsafe condition has been found to exist in particular aircraft, engine, propellers, or appliances installed on aircraft and (2) that condition is likely to exist or develop in other aircraft, engines, propellers, or appliances of the same type design. ADs are substantive regulations issued by the FAA in accordance with Part 39 of the Federal Aviation Regulations (14 CFR Part 39). Once an AD is issued, no person may operate a product to which the AD applies except in accordance with the requirements of that AD.

The process of developing an AD includes analysis and recommendations from the equipment manufacturer and/or designer, an economic analysis of the costs and benefits attributable to the AD, and a period for public comment. The AD must include all the forgoing matters, including all inputs from involved parties. In a situation that requires it, there are provisions for handling such matters expeditiously, and other elements of the AD can be completed later.

[6] Damage tolerance for military aircraft, for example, can include a need to withstand ballistic damage. See Kim, Sheehy, and Lenhardt, 2006, for a comparison of the approaches of three military services: the U.S. Navy, the Canadian Forces, and the U.S. Air Force.

**Figure 1.3
Topical Overview**

**Standard
practices**

**U.S. Air Force's standard practices (MIL-STD-1530C) for fielding durable and
damage-tolerant structures**

Task I.   Design information
Task II. Design analysis and development
          testing
Task III. Full-scale testing
Task IV. Certification and force-
          management development
Task V.   Force management execution

– Air Force policy (AFPD 63-10) requires
  tailoring
– Technical realities necessitate tailoring
– Systems engineering can provide
  tailoring tools

**Tailoring
by system
engineering**

Tools of systems engineering that can tailor practices for hybrid structures

– A "V" framework for systems engineering of durability and damage tolerance
– Synthesis and value-added analysis of engineering opportunities
– Systems-engineering plan for tailoring the general approach
– Additional systems-engineering functions that can facilitate tailoring

Challenges

**Challenges during development and sustainment phases that must be addressed
by a systems-engineering approach**

Technical challenges
  – Multiple expectations
  – Range of possibilities for design and
    causes of damage
  – Variety of damage mechanisms
  – Capabilities of design analyses
  – Capabilities of testing

Programmatic challenges
  – Familiarity with systems
    engineering
  – Time and resources
  – Curiosity about risks

Opportunities

**Opportunities for improving communication and collaboration to
overcome challenges**

– Collaborate on matters of mutual interest
– Transform separations into bonded relationships
– Think, talk, and act with a common language across engineering disciplines
– Focus collaborations on areas with potential challenges for hybrid structures

A way ahead

**A framework for improved collaboration among engineers from systems,
structures, and materials disciplines**

– A value chain for observing and controlling the integrity of aircraft structures
– Process for defining, evaluating, and prioritizing prospective contributions of
  engineering activities to the value chain
– A value-chain approach to stating the business cases for engineering activities
– Criteria for budgeting resources, time, and safety margins to engineering activities

RAND *TR489-1.3*

CHAPTER TWO

The General Air Force Approach to Fielding Durable, Damage-Tolerant Structures

Catastrophic structural failures caused by metal fatigue was recognized as an engineering problem for transportation systems for trains in the middle 1800s (Schutz, 1996). This failure mode continued to be a recognized engineering problem deep into the next century, even as the structures of commercial and military aircraft incorporated new high-strength metals in the 1950s, 1960s, and 1970s. The failures of that period motivated significant advances in the fielding of durable, damage-tolerant structures fabricated from metal components late in the 20th century.

Since the problem of a single fatigue crack causing a catastrophic failure in an otherwise healthy part has been brought under control for metal parts, two important changes have occurred. First, the concept of failure has expanded to include many additional attributes that are of keen interest to operators. These include performance, cost, availability, and reliability. Second, metal parts are no longer the overwhelmingly dominant class of material that is used to fabricate many airframes. The Air Force has recently responded to these changes by updating its approach to aircraft structural integrity. This DoD-approved approach, described in DoD MIL-STD-1530C (2005), defines the standard practices for the Aircraft Structural Integrity Program (ASIP) for U.S. Air Force aircraft.[1] The approach includes five tasks over a structure's life cycle:

- Task I: Design Information
- Task II: Design Analysis and Development Testing
- Task III: Full-Scale Testing
- Task IV: Certification and Force-Management Development
- Task V: Force-Management Execution.

ASIP's task framework is designed to support the understanding, modeling, and management of the technical factors that contribute to failures. For example, materials and structural engineers work on identifying the sciences of failure by researching the potential roles of the physical, chemical, and biological processes that may be involved. Once the dominant process has been identified and described in terms of its driving factors (operating cycles, time, envi-

---

[1] See MIL-STD-1530C, 2005, and NRC, 1997, for a full description of the practices, their purposes, and the value that they add. For background on the technical approach, see Paris, 1961, 1964; Gebman and Paris, 1977, 1979; and Forman, 2002. For information about the history of the evolution of the approach, see Coffin and Tiffany, 1976; National Research Council (NRC), 1997, and Lincoln, 1996, 1997.

ronment, etc.),[2] the next step is to develop engineering models of how process outcomes relate to process inputs, such as operational environments and cumulative use. The ultimate objective of such work is to develop and apply engineering tools that can be used to guide the design, manufacturing, and sustainment of durable, damage-tolerant structures.[3]

Today, the Air Force is using its ASIP task framework to prevent performance, cost, availability, reliability, and (of course) safety failures. Among the key questions that framework asks are those that apply to individual types of failure:

- What constitutes this failure?
- What causes it?
- What are its consequences?
- Does this type of failure have degrees of severity?
- How likely is this failure to occur?
- Can it be prevented?
- Can its consequences be controlled for or mitigated? If so, at what cost?
- What risk levels are associated with this failure?

Finally, a related but more-general question asks what risk management practices would need to be taken to reduce the likelihood that a product would fail for controllable or understandable reasons.

Meanwhile, new hybrid materials and new hybrid structures continue to emerge in many parts of new aircraft. Such materials support numerous subsystems within the airframe, along the airframe's exterior surface, and attached to its exterior surface. Issues of durability and damage tolerance also arise in such applications. Although the ASIP approach does not directly address such applications, its general framework can also be tailored for their engineering. Following is a relatively brief outline of the general approach defined in MIL-STD-1530C.[4]

Design Information (Task I)

Task I includes the development of design specifications, design criteria, and design characteristics:

- specification of the environment and use for which the aircraft is to be designed and associated specification of the aircraft's design service life (Task I-1.1)[5]

---

[2]   Dominant processes with the current era's set of aging aircraft include cracking, corrosion, delamination, and deterioration of adhesive bonds, coatings, and sealants.

[3]   Models, for example, can assess the cumulative damage resulting from operations and environmental exposure. Using such models prospectively during design can forecast outcomes from planned use. Using them retrospectively can assess cumulative damage from known actual use. Such assessments can help guide investments in sustainment (e.g., modifications). They also can help guide decisions about when and how fast to replace a fleet of aircraft.

[4]   This report lists the major tasks to provide the reader an outline of the approach's scope and orientation. See MIL-STD-1530C, 2005, and NRC, 1997, for details.

[5]   Each major task, such as Task I, has a number of subtasks that this report identifies by subtask codes, such as I-1.1. In Chapter Five, specific subtasks are identified by the subtask codes assigned here.

- definition of structural design criteria specifying the nature and extent of loads that the aircraft must be capable of sustaining (I-1.2)
- selection of structural concepts, materials, material fabrication processes, and joining methods (I-1.3).

Task I also includes specifications for processes that will ensure the aircraft's structural integrity:

- a control program for ensuring the structure's continued durability and damage tolerance in the presence of material damage (I-2.1)
- a control program for preventing unacceptable development of corrosion (I-2.2)
- a program for nondestructive inspection of the structure that will preclude the development of dangerous degradation of the structure from all relevant damage mechanisms (I-2.3).

The results of Task I and subsequent tasks are coordinated through a master plan for an individual ASIP (I-3).

Design Analyses and Development Testing (Task II)

Task II includes the characterization of the environment in which the aircraft must operate; the initial testing of materials, components, and assemblies; and the analysis of the aircraft design. This task includes the following activities:

- design development activities
  - loads analysis (Task II-1.1)
  - design-spectra analysis for the service loads (II-1.2)
  - design-spectra analysis for the chemical and thermal environment (II-1.3)
  - testing of allowable loads and environments for materials and structural joints (II-1.4)
  - stress analysis (II-1.5)
  - mass properties determination and analysis (II-1.6).
- design analysis and evaluation activities
  - vibration analysis (II-2.1)
  - aeroelastic and aeroservoelastic analysis (II-2.2)
  - sonic fatigue analysis (II-2.3)
  - durability analysis (II-2.4)
  - damage tolerance analysis (II-2.5)
  - survivability analysis (II-2.6)
  - corrosion assessment (II-2.7)
  - initial risk analysis (II-2.8).
- design-development tests
  - verify the results of design analyses and evaluations of the design's major elements (II-3.1)
  - discover design deficiencies in the design's major elements (II-3.2).

Finally, the capability to perform nondestructive inspections during production is assessed and verified (II-4).

Full-Scale Testing (Task III)

Task III consists of flight and laboratory tests of the aircraft structure to assist in determining the structural adequacy of the analysis and design, including

- static tests (Task III-1)
- first-flight verification ground tests (III-2)
- flight tests (III-3)
- durability tests (III-4)
- damage-tolerance tests (III-5)
- climatic tests (III-6)
- interpretation and evaluation of test results (III-7).

Certification and Force-Management Development (Task IV)

Task IV includes the analysis that (1) defines the flight envelope for safe operation of the aircraft's structure and (2) provides the basis for certification of the aircraft's structure.[6] The subtasks include

- strength summary and operating restrictions (Task IV-1.1)
- certification analyses (IV-1.2).

Task IV also includes the development of the processes and procedures that will be used to manage operations and sustainment of the fleet (inspections, maintenance, modifications, damage assessments, risk analysis, etc.) when its aircraft enter the inventory. The subtasks include

- load and environmental spectra survey development (IV-2.1)
- individual aircraft tracking program development (IV-2.2)
- rotorcraft dynamic component tracking program development (IV-2.3).

These processes and procedures produce and maintain the force structural maintenance plan (IV-3).

---

6   The determination of the flight envelope involves much iteration, starting during the design process and continuing with trade-offs occurring throughout the development of the aircraft, including structural testing and modifications that may occur following tests.

Force-Management Execution (Task V)

Task V executes the processes and procedures developed under Task IV to ensure structural integrity throughout the life of each individual aircraft. This task may involve revisiting elements of earlier tasks, particularly if the service-life requirement is extended or if the aircraft is modified:

- load and environmental spectra survey (Task V-1.1)
- individual aircraft tracking program (V-1.2)
- rotorcraft dynamic component tracking program (V-1.3).

Products generated during Task V include

- ASIP manual (V-2.1)
- aircraft structural records (V-2.2)
- force management updates (V-2.3).

Recertification may be required because of changes and/or aging (V-3).

CHAPTER THREE

Systems Engineering Tools That Can Help Tailor the General Approach to Hybrid Structures

The previous chapter described a comprehensive framework of tasks that can be implemented to ensure the durability and damage tolerance of aircraft structures, including hybrids. However, that general approach requires tailoring, both for Air Force policy and technical reasons.[1] Each acquisition program must tailor its own ASIP to satisfy the requirements for its system's airframe. Thus, the level and composition of investments in engineering activities for each of the five ASIP tasks can vary across weapon systems because of differences in requirements. Technically, tailoring is necessary because, while the general approach was initially defined and implemented to manage structural fatigue of metal components, hybrid structures have a variety of failure modes that require new technical approaches to implementation. The general approach does, however, provide a broadly applicable framework. Moreover, it already includes many of the tasks and subtasks that are important to all structural components, independent of the material from which they may be fabricated.

For hybrid structures in airframes, this chapter identifies tools of systems engineering that can be used to help implement and, where necessary, tailor the Air Force's general approach. Figure 3.1 illustrates the general relationship between systems-engineering activities and functional-engineering activities. Generally, a program's project managers base the level and composition of investment in engineering activities on information from both functional engineers and systems engineers. The functional engineers are responsible for engineering the specific components and the structure of the system, which is an important, but not the only, element. In contrast, the systems engineers are responsible for the integrity of the total system, which they evaluate by testing, operating, and monitoring components and the system as a whole.

Many systems-engineering tools can contribute to effective implementation of the general approach to hybrid structures by providing a rigorous process that

1. identifies the tasks and subtasks that need implementation and that may require tailoring
2. facilitates the synthesis of engineering opportunities for implementing, modifying, replacing, and/or adding tasks or subtasks
3. evaluates each opportunity's comparative potential for adding customer-critical value
4. proposes alternative plans for implementation.

---

[1] See Air Force Policy Directive 63-10, 1997, and Air Force Instruction 63-1001, 2002, for policy and instructions regarding the tailoring and implementation of the general approach.

Figure 3.1
Systems Engineering Contributing to Durable, Damage-Tolerant Structures

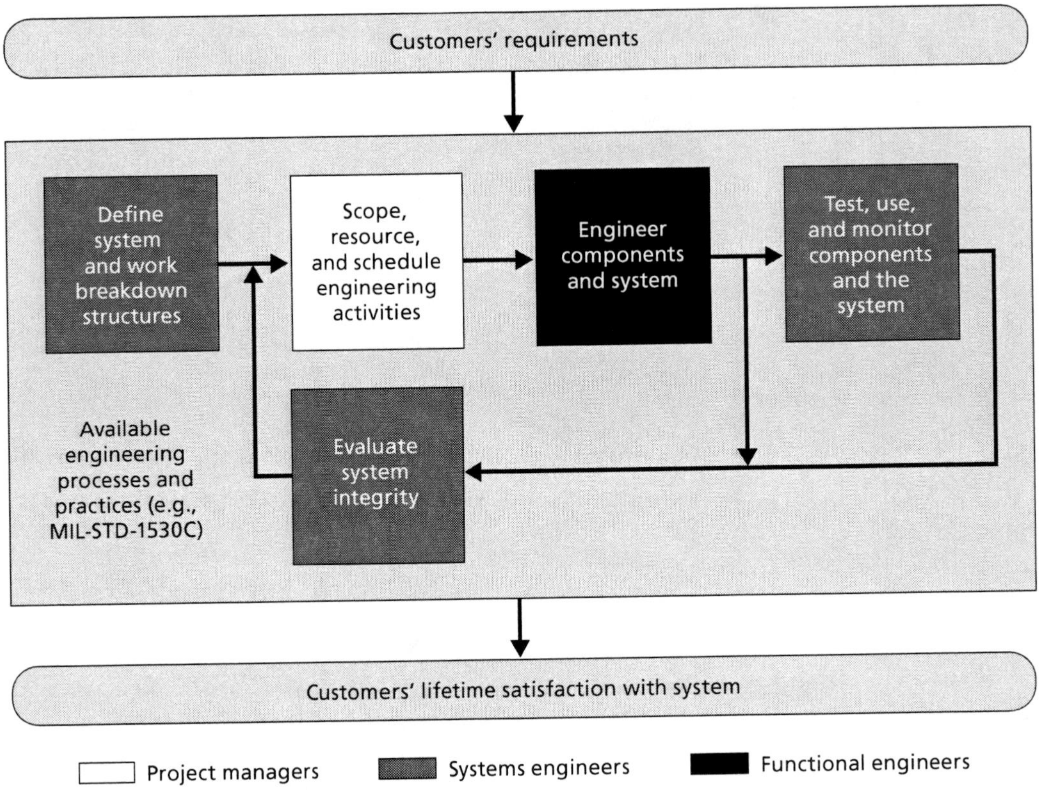

RAND TR489-3.1

Development of such a process could be guided with the assistance of the systems-engineering activities and tools described in the rest of this chapter.

Development of a "V" Framework for Engineering for Durability and Damage Tolerance

Three sets of systems-engineering tasks form a framework for synthesizing and evaluating prospective engineering opportunities related to durability and damage tolerance.

Flow-Down of Durability and Damage-Tolerance Requirements

In this set of tasks, systems engineers begin by examining the contractual requirements for the performance and reliability of the overall system (including its durability and damage tolerance). The engineers then translate these into equivalent requirements for each of the system's elements, defining performance and reliability requirements for each so that the overall system can comply with its own requirements.[2]

---

2   Structural engineers must also look ahead, beyond the development contract to address lessons from such experiences as those in current operations: new damage modes from hostile fire, changes in aircraft missions, changes in aircraft loads, and damage from unscheduled maintenance in forward-deployed sites. Such matters are addressed subsequently, under proactive identification of risks.

Building Customer-Critical Value Through Increasing Durability and Damage Tolerance

Customer-critical value is reflected in the contractual requirements for durability and damage tolerance. As the system is built up from its smallest pieces to successively higher levels of assembly, customer-critical value is added at each level in terms of durability and damage tolerance.[3]

Figure 3.2 illustrates the "V" framework, which provides a way of thinking about this flow-down of durability requirements (left side of the V) and buildup of customer-critical value (right side of the V). The top of the V is the customer interface. At top left is where the customer's top-level requirements are stated. At top right is the customer's experience with the final product. The left side of the V deals with breaking the product down to progressively smaller pieces; this is decomposition. The right side deals with building the product up to progressively larger assemblies, with appropriate validation and verification at each level of assembly.

Horizontal paths linking items listed on the right and left sides of the V identify opportunities for validation and verification at each level of decomposition and for each component at that level, ultimately including the entire aircraft and its maintenance system, at the top of the V.

Throughout both the flow-down and buildup phases, engineers must be attentive to both the stated requirements and the customer's final experience. The V model is a tool for seeing and managing the big picture.

Observing the Buildup of Value

Because customer satisfaction (or lack thereof) may not be apparent until much later in the life cycle for metrics related to durability and damage tolerance, observations about the buildup of value during the design and manufacturing phases are essential. Although direct observations often are impossible, except when accelerated life testing and residual-strength testing are options, signs of durability and damage tolerance can often be inferred from observations of processes and design characteristics. For example, at the part and material levels, it is not possible to observe the buildup of customer-perceived value directly during the design and manufacturing phases. It is possible, however, to observe the processes that are used to provide reasonable assurance of suitable durability and damage tolerance. Does the material have an established track record? If the material is new, is there a reasonably thorough program of testing and prototyping that precedes a decision to go forward with the material? Is there a backup material ready for use, just in case?

Synthesis and Value-Added Analysis of Engineering Opportunities

The second major systems-engineering activity has three sets of tasks that contribute to the synthesis and evaluation of prospective engineering opportunities related to durability and damage tolerance.

---

[3] Over time, facets of product quality that are not articulated in contractual requirements also reflect value. In some situations, requirements might be amended to address such matters. In other situations, the value may not be easy to capture in a procurement contract. For example, a given design detail may have a high risk of causing a problem later in the structure's service life, but this would not become evident during any of the qualification and acceptance testing. Recognizing, accounting for, and evaluating the value such quality enhancements add can be part of (or an adjunct to) the analysis of the buildup of contractually required value.

Figure 3.2
Systems-Engineering Areas Involved in the Decomposition of a System and the Buildup of System Value: The V Model

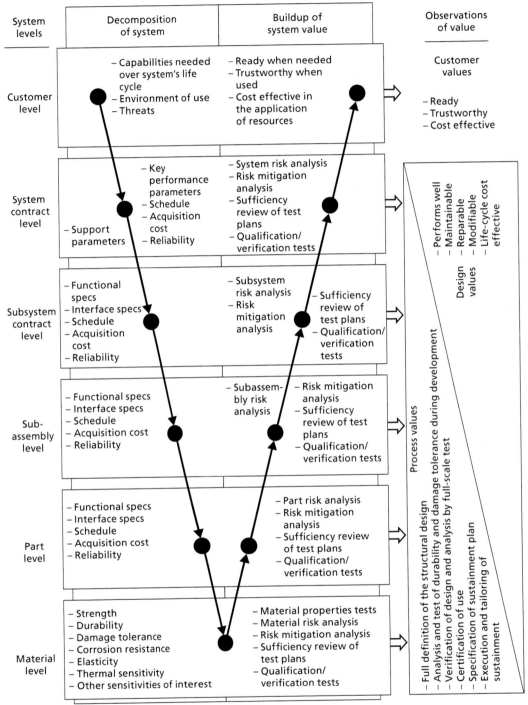

Proactive Identification of Risks

The task of proactively identifying risks may be among the most difficult and the most critical tasks related to tailoring the general approach. Here, for example, is where matters that the contractual requirements do not cover need to be addressed.[4]

Engineering-Opportunity Identification

For each identified risk, this task synthesizes, evaluates, and selects the most promising engineering alternatives for achieving one or more of the following outcomes:

1. a reduction of the likelihood that the risk will occur
2. a reduction of the nature and/or extent of the damage that would occur
3. an increase in the structure's ability to operate satisfactorily until the damage would be detected
4. an increase in the likelihood that the damage would be detected within an acceptable period of time
5. improvement of the ease with which the damage would be repaired in a sufficient manner

Critical Analysis of Value Added by Engineering Opportunities

For each of the most promising engineering alternatives, this task

1. evaluates the comparative potential for the engineering alternative to add customer-critical value
2. estimates the schedule and resource requirements for executing the alternative
3. compares the costs and benefits of this alternative with those of other alternatives.

Systems-Engineering Plans for Tailoring the General Approach

The third major systems-engineering activity has a set of four tasks that create and evaluate alternative courses of action for the engineering effort. The first three look at plans with increasing levels of risk; the fourth compares these alternatives:

1. lowest risk alternative—leaves schedule and resources unconstrained and includes the least-cost set of engineering activities that is consistent with a low-risk implementation[5]
2. intermediate risk alternative—constrains both the schedule and resources and assumes the set of engineering activities consistent with best-case expectations about the time and resources available for the engineering effort[6]

---

[4] Structural engineers, for example, need to consider lessons from ongoing experiences that may indicate future risks that current contractual requirements do not address.

[5] For example, a *least-risk plan* might incorporate most of the provisions of MIL-STD-1530C, including thorough verification testing at each level of assembly for durability and damage tolerance.

[6] For example, an *intermediate-risk plan* might compensate for a compressed schedule by specifying relatively mature materials and structural concepts.

3. highest risk alternative—constrains both the schedule and resources and assumes the set of engineering activities consistent with worst-case expectations about the time and resources available for the engineering effort[7]

4. evaluation of alternatives—evaluates the prospective costs and benefits of the three alternatives for tailoring the general approach to fielding durable, damage-tolerant structures.

Additional Systems-Engineering Functions That Can Facilitate Tailoring

The following additional systems-engineering functions can facilitate the tailoring process:

- flow-down of requirements for performance and reliability; this includes a corresponding flow-down allocation of responsibilities and resources
- development of a risk-management plan
- scoping and scheduling of engineering activities
- determination of the need for qualification and/or verification tests at each level of assembly
- sufficiency reviews for test plans at each level of assembly
- development and review of interface control documents
- independent analysis of technical alternatives when problems are encountered
- independent, objective, and balanced assessments of risks associated with alternative courses of action and the definition of risk-mitigation portfolios required for each course of action that might be selected.

In general, a strong systems-engineering effort starts with system conceptualization. Such an effort deals with each revision of the work plan for the remaining work in a realistic manner that right sizes the remaining effort always. It also provides for proactively managing risks continuously.[8]

---

7 For example, a *highest-risk plan* might delete testing at many levels of assembly to match the anticipated schedule for a program.

8 Evidence of such practices can be found in the previously cited programs: the naval nuclear propulsion program, the ballistic missile programs of the 1950s and 1960s, and the continental air defense program of the 1950s and 1960s.

CHAPTER FOUR

Considerations the Systems-Engineering Approach Must Address

Technical and/or programmatic challenges may impede the process of tailoring the general approach.

Technical Challenges

Technical challenges during the development and sustainment phases include a combination of multiple expectations, a wide range of technical possibilities, a large variety of potential damage mechanisms, the limitations of design analysis for evaluating damage mechanisms, and the limitations of testing for damage mechanisms.

Multiple Expectations

In addition to serving its load-carrying purposes over a design service life, structures for military systems, such as aircraft, must also meet requirements for such characteristics as the structure's ability to survive in the face of hostile fire and harsh operating conditions and its suitability for maintenance, inspection, repair, and modification. Such characteristics must exist for nonstructural material as well as structural components.

Wide Range of Technical Possibilities for Hybrid Structures

Because a hybrid material marries substances with different strengths and weaknesses to form a combination that has strong features suitable across a range of uses, the variety of possibilities is very large. Hybrid structures are being fabricated from combinations of such materials as metals, nonmetals, other hybrid materials fabricated from metal and nonmetal ingredients, and other hybrid materials fabricated solely from nonmetal ingredients.

These materials can have interesting mixes of properties. For example, an external surface may be wear resistant, but because it is brittle and thin-layered, it relies on the next structural layer for other durability attributes. In another example, a substrate may be very highly resistant to the low cycle fatigue that dominates the fatigue durability of the neighboring structure.

Design Possibilities. Thus far, most hybrid materials have been manufactured from two materials. The two dominant reinforcing styles for aerospace parts have been honeycomb and fibers. Various combinations of honeycomb cores and face sheets have emerged, as have various fiber and matrix combinations. In addition to the variety of combinations, the technologies of the constituent materials continue to advance, offering improved mechanical properties, weight, and cost.

Possibilities for Incurring Damage. Because of the complexities of manufacturing hybrid materials and the structures made from them, there are also many more opportunities for damage to occur over the life cycle of parts and hybrid structural assemblies, such as the following:

- material production—fabrication of hybrid materials and assembly of hybrid structures
- assembly—fabrication of small assemblies, joining of small assemblies, assembly of intermediate assemblies, joining of intermediate assemblies, and final assembly of major assemblies
- surface finishing—sealing and coating
- use—wear, fatigue, corrosion, deterioration, impact during ground servicing, hail, bird strikes, and combat damage
- repair—stripping, repairing, and coating
- maintenance—stripping, surface inspection, opening, interior inspection, repair, sealant repair, interior coating repair, closing, and coating
- reinforcement and modification—stripping, opening, attaching, sealing, closing, and coating
- material replacement—stripping, opening, cutting, removing, installing, attaching, sealing, closing, and coating.

Many Potential Damage Mechanisms

Experience with hybrid materials and hybrid structures has identified a variety of damage mechanisms, as the following subsections describe.

Main Mechanisms, Thus Far. The main types of damage to or defects in hybrid structures have included degradation of load-transfer capacity at fasteners, voids, bond failures, delamination, holes, punctures, and cracks. Much of the damage is due to discrete sources, such as impacts, lightning strikes, and handling, rather than progressive growth caused by fatigue.

Mechanical Fatigue. Fatigue is not generally a significant damage mechanism with many hybrid structures that meet impact damage-tolerance requirements.

Deterioration of Joints. Joints are normally the dominant issue for durability and damage tolerance, whether the structure is bonded, welded, bolted, or otherwise joined. Joints may deteriorate over time for any one of a number of reasons, including environment, chemistry, electrochemistry, thermal, and mechanical mechanisms. For example, load transfer at fastener holes can be problematic for hybrid materials, as can fretting at the faying surfaces, where parts make contact in a joint.

Serial-Failure Mechanisms. As an example of this mechanism, an especially difficult maintenance issue occurs when perforation of a face sheet allows hydraulic fluids, water, or other liquids to move into the honeycomb core. Material deterioration may result from corrosion or chemical reactions.

Potential Future Mechanisms. Potential degradation mechanisms to monitor in the future include cracking due to mechanical or thermal stresses; growth of impact damage under fatigue loading; growth of manufacturing-induced damage, especially from fastener installation; and development of corrosion in adjacent dissimilar materials.

Limitations of Design Analyses of Damage Mechanisms

Design analysis tools are being developed and design guidance documents are being developed as new materials emerge and as their failure mechanisms are becoming known.[1]

Design Guidance. Design guidance for hybrid parts and hybrid structures continues to evolve. It addresses the selection, design, and analysis of hybrid structures and includes considerations of static ultimate strength, durability, and damage tolerance and the effects of structural degradation mechanisms, such as impact and humidity (or fluid) exposure.

Models of Failure Mechanisms. The limitations of design analysis methods are reflected in the challenges that researchers have encountered in attempting to predict damage initiation and damage growth in hybrid structures.[2]

Limits of Testing

Ideally, each identified damage mechanism would be tested to verify structural designs, with evaluations at levels from coupon to full scale. The final step may include a full-scale component fatigue test on an impact-damaged structure.[3] Accelerated life testing, however, is not a technical possibility for many damage mechanisms. Thus, in practice, a subset of damage mechanisms is addressed, such as

Impact Damage Testing. To verify impact tolerance, the structure is subjected to a low-velocity impact prior to the fatigue testing to substantiate inspection intervals and performance for the life of the structure under barely visible impact damage criteria.

Humidity (or Fluid) Exposure Testing. Design properties based on coupon tests are typically generated in a fully saturated humidity condition (85-percent relative humidity) and over a range of high temperatures.

Programmatic Challenges

Major programmatic challenges—including lack of familiarity with systems engineering, tight development schedules, limited resources, and lack of a healthy curiosity about risks—can impede the establishment of a sufficiently strong systems-engineering effort.

Limited Familiarity with the Mission and Roles of Systems Engineering

Engineering can be viewed as the technical work done in wisely directing the application of scarce resources to important needs of society. This definition assumes that the necessary conditions for wise application of resources include

- *consistency* with society's established bodies of knowledge
  - the enduring hard facts from the physical and mathematical sciences
  - the best practices of the management sciences
  - the rules, regulations, and laws from the political sciences

---

[1]   For composite materials, see Military Handbook (MIL-HDBK) 17/1F, 2002; MIL-HDBK-17/2F, 2002; MIL-HDBK-17/3F, 2002; MIL-HDBK-17/4A, 2002; and MIL-HDBK-17/5, 2002.

[2]   For corrosion of aerospace metals, especially intergranular corrosion, progress has been slow in developing working models of the failure process, see Defense Science Board, 2004.

[3]   An entire structure may not be designed to be impact resistant, however.

- applications from other relevant bodies of knowledge, including the behavioral and organizational sciences
- *cost effectiveness* of a given resource in comparison to viable alternatives
- *risk appropriateness* in view of recognizable uncertainties and society's tolerance for unintended adverse consequences
- *due diligence* in the effective application of knowledge, including objective, independent, and balanced research and analysis.

Because a system can be viewed as a connected set of elements that, as a whole, aims to realize the wise application of scarce resources to important needs of society over a defined life cycle, systems engineering can be viewed in the following way: ***Systems engineering is the technical work done in wisely directing the technical application of scarce resources to a connected set of elements that, as a whole, aims to realize the wise application of scarce resources to important needs of society over a defined life cycle. It is assumed that necessary conditions for wise application of resources include matters of consistency, cost effectiveness, risk appropriateness, and due diligence as described above.***

Thus, the objective of effective systems engineering is the cost-effective orchestration of the total engineering effort across all disciplines and throughout a system's life cycle, from concept formulation through final disposal of the system. The value of a strong systems-engineering effort rises rapidly with the system's size, the system's technical complexity, the number of engineering disciplines involved, the technical complexity of any of the system's materials, the technical complexity of any of the system's software, the technical complexity of interfaces among the system's subsystems, the technical complexity of any interfaces with other systems, the extent of any reliance on emerging technologies, and the system's remaining service life.

Limited Time and Resources

During the 1950s, 1960s, and 1970s, the aircraft industry and its customers learned the consequences of proceeding too quickly with superior-strength materials that later revealed serious weaknesses in durability and damage tolerance.[4]

Limited Curiosity About Risks

If time and/or resources are very tight, curiosity about how structural components can degrade and fail could be limited. Additional factors may include a desire to minimize any concerns that prospective customers may have about the level of risk for a new product. Such concerns may also contribute to an aversion to engineers thinking deeply about all the potential ways in which a new part might fail to deliver necessary durability and damage tolerance.

---

[4] As noted in Chapter Two, new, high-strength metals were later found to have serious limitations in terms of durability and damage tolerance.

CHAPTER FIVE

Opportunities for Collaboration Among Systems, Structural, and Materials Engineers

For something as complex as a hybrid structure for an airframe, cost-effective engineering of the necessary levels of durability and damage tolerance will require systems, materials, and structural engineers to speak a common language and think and act in concert. Such collaboration must start with an effective articulation of the business case for the necessary engineering activities over the structure's life cycle. It must continue with effective revisions of the business case for necessary engineering activities as the design, its use, and its operating environment evolve over the vehicle's life cycle.

This chapter describes the elements of this sort of collaboration and what the team members can do to benefit from each.

Collaborate on Matters of Mutual Interest

Engineers in all three disciplines under discussion share two fundamental interests: proactively and continuously managing risks and always right-sizing tasks. These matters are also of interest to project managers, system managers, and customers because of their effects on how schedule, cost, and performance balance one another over the system's service life. Budget and schedule pressures, however, can create tensions that may weaken adherence to such process principles. Thus, materials engineers, structural engineers, and systems engineers need to collaborate in making the best business case for the right course of action.[1]

Transform Current Separations into Bonded Relationships

Although attitudes and choices about schedules and resources and curiosity about risks affect all three engineering disciplines (systems, structures, and materials), these disciplines often continue to be separated by language, focus, principles, and the types of actions they involve. One way to strengthen the bond between systems and structural engineering, as well as that between structural and materials engineering, would be to tackle these four differences and reformulate them into a four-part epoxy by, for example, addressing each as follows:

---

[1] The business case needs to take a broad view that includes considerations of maintenance, life-cycle cost, and any component testing that needs to be done at the joint level.

- Develop a *common language* for critical matters that is clear, concise, and compelling.
- *Focus* on customer values over the life cycle of the system.
- Adopt a set of value-based process *principles* to guide the allocation of time and resources and to guide the development of reliability and durability budgets, performance-error budgets, and margins of safety, both at the system level and throughout the system break-down structure.[2]
- Recognize and apply the type of *action* that circumstances require by adjusting quickly in a coordinated way across disciplines.

Focus Collaboration on Areas with Potential Challenges for Hybrid Structures

Some tasks and subtasks within the general approach are in areas in which hybrid structures may encounter challenges that are very different from those for metal structures. Because the significance of such differences may depend on the nature and characteristics of a particular structure, the following observations illustrate the types of matters that might arise in tailoring the general approach to specific hybrid structures.

Design Information (Task I of MIL-STD-1530C)

Selection of materials (Task I-1.3) may pose special challenges when the nature and extent of advances in materials technology create the possibility of new damage mechanisms or new ways of initiating old mechanisms. Also, the integrity-control programs (Task I-2) may need new approaches and methods to adjust to the different natures of new materials.

Design Analyses and Development Testing (Task II of MIL-STD-1530C)

The testing of structural joints (Task II-1.4) may also pose new challenges. Wherever different materials come into contact in a joint, it is necessary to explore potential vulnerabilities due to chemical, electro-chemical, biological, and mechanical interactions. Corrosion in metal joints is a classic example. Also, the initial risk analysis (Task II-2.8) may have to address a broader range of risks. And assessing and verifying capabilities for nondestructive inspection (Task II-4) of all the different types of materials, damage mechanisms, structural details, and joints could pose a new set of challenges.

Full-Scale Testing (Task III of MIL-STD-1530C)

Full-scale damage-tolerance testing (Task III-5) may prove challenging because of the variety of places in which damage may occur, the various possible causes of damage, and the variety of types of damage.

---

2   One example of a margin of safety for a structural assembly would be a requirement to sustain 80 percent of the design's limit load even after the failure of any single element of the structure. A higher margin of safety would be to require a capacity to sustain 100 percent of limit load. A different example of a margin of safety would be requiring two lifetimes of durability testing instead of only one.

Certification and Force-Management Development (Task IV of MIL-STD-1530C)

Load and environmental spectrum survey development (Task IV-2.1) might have to be adapted to take into account the differences between how metals and various types of hybrids react to environmental exposure.

Force-Management Execution (Task V of MIL-STD-1530C)

Tracking instrumentation for individual aircraft (Task V-1.2) might have to be expanded to monitor additional aspects of structural health, focusing on new as well as traditional materials. Also, aircraft structural records (Task V-2.2) might have to be expanded to capture information about incidents that could contribute to hidden degradation of material. Examples might include impacts, lightning, and liquid spills (water, lubricants, etc.).

CHAPTER SIX

Framework for Strong Collaboration Among Systems, Structural, and Materials Engineers

Figure 6.1 offers a framework for facilitating strong collaborations by engineers from the systems, structures, and materials disciplines. The figure depicts the following elements:

- *A value chain for the integrity of aircraft structures:* Structural integrity is adherence to a value chain (described below) that includes customer values, design values, and process values, all of which help drive the tailored engineering approach.
- *A process for defining, evaluating, and prioritizing prospective contributions of engineering activities to the value chain:* Effective engineering of complex systems requires a concept-development activity that is linked to a development-planning activity in which alternative architectures for a product and alternative engineering approaches are explored in an iterative process that produces the best architecture and the best engineering plan for satisfying customer values. Once the product architecture has been frozen, a detailed engineering plan can be developed.
- *Template for building business-case statements for engineering activities:* An effective statement of the business case for an engineering activity shows how that activity adds value to each link in the value chain.
- *Criteria for applying time, resources, and safety margins to engineering activities:* Using consistent value-chain-based criteria can support the effective application of time, resources, and safety margins across engineering activities and over a product's lifetime.

Consider, for example, the dilemma about whether to test a new material in the laboratory or to conduct a full-scale test later that simulates the intended operational use. This dilemma could be analyzed by considering the prospective contributions to the illustrative code of values. If skipping the laboratory tests incurs a very high risk and if the cost of changing materials after a full-scale test is very high because a large development program would experience a very costly delay, the customer value of "cost-effective use of resources" would seem to be at an inordinately high risk, unless there is some overriding consideration, such as an urgent need to field the system.

A Value Chain for the Integrity of Aircraft Structures

Structural integrity relies on adherence to a code of values that can be decomposed into a three-level system of values. At the top level are the customers who will use the aircraft structure. At the middle level are the necessary design characteristics of the aircraft structure that

29

Figure 6.1
A Framework for Collaboration by Materials, Structures, and Systems Engineers

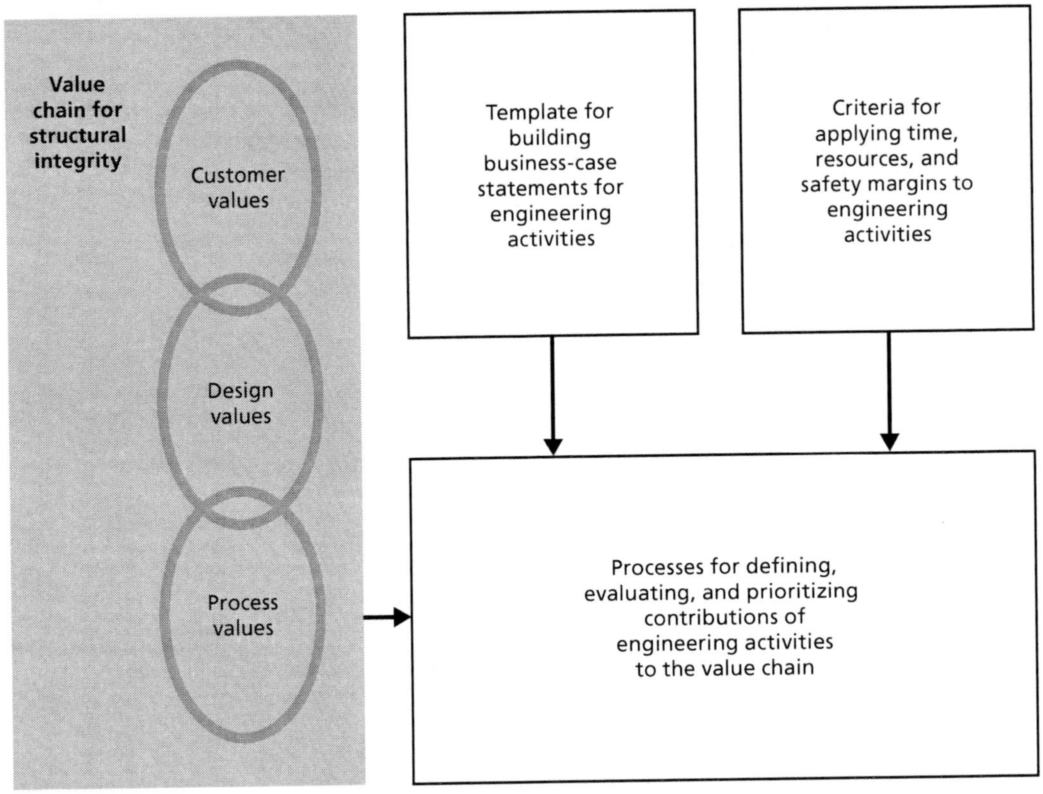

RAND *TR489-6.1*

will serve the customers. At the bottom level are the processes that will develop, produce, and sustain the designed characteristics.

Customer Code of Values

Most customers, facing a scarcity of resources, desire a system, product, or service that provides a lifetime of service and that is ready when needed, trustworthy when used, and that uses resources cost-effectively. Such customer interests constitute a *customer code of values*.

Design Code of Values for an Airframe Structure

To satisfy such customer interests, an aircraft manufacturer would want to deliver an airframe structure that, for example,

- performs well: It satisfies its functional performance specifications.
- is maintainable: Its maintenance costs and downtime are reasonable.
- is repairable: Its repair costs and downtime are reasonable, and repairs satisfy reasonable requirements for durability and damage tolerance.
- is modifiable: The costs and downtime for modifying the structure are reasonable, and such modifications satisfy reasonable requirements for durability and damage tolerance.
- is cost-effective over the life cycle: The lifetime costs for acquisition and service are lower than those for competing design alternatives.

Such a set of design interests constitutes a *design code of values*.

Process Code of Values

The Air Force's general approach to fielding durable, damage-tolerant structures provides a framework that embodies a code of process values for guiding ASIPs. The Air Force's general approach aims to establish and sustain aircraft structural integrity for a specified range of uses through adherence to the following set of process values (and associated tasks):

- full definition of the structural design (Task I)
- analysis and test of durability and damage tolerance during development (Task II)
- verification of design and analysis by full-scale test (Task III)
- certification of use (Task IV-1)
- specification of sustainment plan (Task IV-2)
- execution and tailoring of the sustainment plan (Task V).

Process for Defining, Evaluating, and Prioritizing Prospective Engineering Contributions to the Value Chain

System integrity is built through adherence to the value chain. Adherence starts with the development of a detailed engineering plan that clearly defines the goal, approach, and objectives.

- **Goal:** a cost-effective orchestration of the total engineering effort across all disciplines and throughout the system's life cycle, from concept formulation through final disposal of the system.
- **Approach:** includes a systems-engineering framework for tailoring processes and practices that includes (1) the flow-down of requirements, schedules, and resources; (2) an observable buildup of system value; and (3) the identification of each engineering activity's contribution to the value chain.
- **Objectives:**
  - a sound systems framework including (1) a common language, focus, and principles; (2) clear definitions for functions and interfaces; and (3) a coherent plan for integration of efforts
  - business cases that have been stated effectively for each engineering activity, including systems and functional engineering activities
  - value-chain-guided evaluations of business cases for each engineering activity
  - engineering resources that are sufficient and efficiently used
  - a balanced portfolio of engineering activities across engineering disciplines and over the system's life cycle.

The capstone documents for the plan are preparation of a sustainment-engineering requirements plan (SERP) during the sustainment phase and preparation of an SEMP during both the acquisition and sustainment phases. During the sustainment phase, the SEMP reflects the parts of the SERP to which resources are actually allocated.

Responsibility for Adherence to the Value Chain

Systems-engineering work is responsible for the cost-effective orchestration of the total engineering effort across all parts of the system, across all disciplines, and throughout a system's life cycle, from concept formulation through final disposal of the system. This includes the flow-down of responsibilities, authority, and resources. It also includes the buildup of value, starting with engineering processes, moving up to design characteristics, and finishing with customer value. Thus, systems-engineering work provides the central nervous system for integrating the total engineering effort in ways that assure satisfaction of customer values.

In some cases, the systems-engineering work is inherent in the engineering practices that have evolved within a company for a particular product line that has met with great success. Relatively few engineers may actually hold the title of *systems engineer*.

In another case, a company may be seeking to enter a new business area. Here, the investments in systems-engineering positions may be more noticeable because the company is just beginning to sort out how it will orchestrate an effective engineering process for a new class of products.

Either way, the increased use of hybrid materials and hybrid structures creates a need for systems-engineering work, whether or not it is labeled as such. For example, when a new type of joint is to be used to join two different hybrid materials, systems engineering should review how much testing should be done at each level of assembly.

Allocation of Engineering Resources

The level and composition of resources allocated to engineering activities are established by a process of defining, evaluating, and prioritizing prospective contributions of engineering activities to the value chain. Although systems engineers are responsible for orchestrating that process, functional-area engineers (e.g., structural engineers and material engineers) play a fundamental role in defining prospective activities for their areas of responsibility. They need to work closely with the system-level engineers as an integrated team. The system-level engineers are key to helping provide independent assessments of risks across parts of the system, across disciplines, and over time.

Commonality in Language, Focus, Principles, and Actions

Functional-area engineers must make business cases for their perceived needs (resources and time), and systems engineers need to understand both the technical and business implications of such business cases. Key to the engineering effectiveness of the functional areas is the integrity of information flow in two directions:

- across systems and functional engineering disciplines: effective communications from functional areas to the systems engineers
- vertically: the systems engineer's ability to carry information from the functional areas to higher levels of responsibility.

Each of these two factors can benefit from the clarity and conciseness of communication that can be facilitated by common language, focus, principles, and actions.

Effectively Stating the Business Case

In the competition to win contracts and in the subsequent competitions to secure and retain the time and resources to engineer the end product's elements, there is a continuing need to articulate the business case effectively for each engineering activity. Similarly, there is a continuing need to articulate the business case for each safety margin built into the design. What one functional area considers to be a margin of safety, others will consider to be an opportunity cost. To avert inappropriate adjustments of margins of safety in the interests of reducing weight, cost, or some other metric, each functional area must articulate its own business cases effectively.

The Business Case for Systems Engineering

Even the business cases for systems engineering activities need continuous and effective rearticulation over the product's life cycle, from conceptualization through disposal. These cases should specify how each systems-engineering activity relates to the links in the value chain and how the level and composition of resources and the schedule for the activity influence the value chain.[1]

Business Case for an Engineering Activity

To divide time and resources effectively across the system elements and multiple disciplines, it is in the best interest of the overall project for each functional area to communicate its risks and opportunities effectively, in an objective and balanced manner.

After a functional area has recommended a course of action for its own engineering activities, its engineers should work with systems-level engineers as an integrated team to develop a written business case. The remainder of this section addresses the elements of such a business case for a recommended course of action and how to address alternatives.

Case for Recommended Course of Action

The business case for a functional area's recommended course of action could address the following:

- how the activity is relevant to the value chain
- the time and resources required (including a work breakdown structure and a schedule)
- the planned margins of safety, including their supporting rationales
- a risk survey that includes the exhaustive identification of potential risks
- a risk analysis
- any risk mitigation plans.

Alternative Courses of Action

To meet business constraints or satisfy changes in customer requirements, alternative courses of action may need to be identified to support trade studies within and across parts of the system, across disciplines, and across time. A business case for a functional area can help inform such trade studies by identifying alternative courses of action for its area.

---

[1]   Although there is interest in demonstrating the benefit of systems-engineering work in terms of cost savings or some other quantifiable metric, attaining such objectives remains illusive. As an analogy, consider the job of building a simple residence. One approach is to hire an architect to draw a plan and then hire a contractor. A second approach is to hire a contractor and save the cost of paying an architect. How would you go about demonstrating the cost-savings advantage of either approach? Would you rather analyze the circumstances more deeply, then make an informed judgment? What if the contractor were a trusted party who recently built a house just like the one you want? Suppose instead that you are new to the community and do not know anyone; a little analysis plus an informed judgment might produce the best course of action.

Analysis of Different Courses of Action

Each functional area could provide a comparative analysis of its recommended and alternative courses of action. This will help inform subsequent resource and schedule decisions about the sensitivity of the value chain to deviations from the recommended course of action. Such analyses could show potential effects on system-level outcomes.

Criteria for Budgeting Time, Resources, and Safety Margins for Engineering Activities

A consistent set of criteria for budgeting time, resources, and safety margins for engineering activities could be tied to the value chain (customer values, design values, and process values) and the principles of systems engineering, such as always right-sizing remaining efforts and proactively and continuously managing risks.

The testing-dilemma problem described previously can help illustrate these points. When the introduction of a new material brings significant risks, adequate testing at each level of assembly can be crucial. Cutting corners on testing can be very risky. Right-sizing a bid to include adequate testing can be very important in such circumstances.

On the other hand, if the bid was reduced in response to competitive pressures or for other reasons, the proactive risk-management path might call for testing as early as meaningfully possible, rather than simply hoping for the best and reacting later, only if difficulties arise.

CHAPTER SEVEN

Conclusion

Materials engineering is entering a phase when the variety of possibilities and the rate of arrival of seemingly newer and better materials may be reaching far beyond experience. In particular, this report has identified a number of key findings and conclusions:

- Hybrid structures are becoming increasingly complex.
- Systems engineering is already making valuable contributions to many complex systems.
- Materials and structural engineering could link up more effectively with systems engineering.
- Effective management of aging hybrid structures could benefit from a strong systems-engineering effort and a comprehensive, integrated life-cycle approach.

As industry and operators are tailoring and implementing the Air Force's general approach, this may be a good opportunity to consider how well material engineers, structural engineers, and systems engineers are performing as a team in ensuring the durability and damage tolerance of hybrid structures for airframes over their life cycles. The collaborative, cross-discipline framework this report has described may be a useful basis for considering and continuously improving the team performance of material, structural, and systems engineers who are responsible for ensuring the durability and damage tolerance of hybrid structures over an airframe's life cycle.

Bibliography

Air Force Instruction 63-1001, *Aircraft Structural Integrity Program*, April 18, 2002.

Air Force Policy Directive, 63-10, *Aircraft Structural Integrity*, November 1, 1997.

Air Force Space and Missile Systems Center, *Systems Engineering Primer and Handbook: Concepts, Processes, and Techniques,* El Segundo, Calif.: Los Angeles AFB, 2004.

Barker, Joel Arthur, *Paradigms, The Business of Discovering the Future,* New York: Harper Collins, 1993.

Baum, Claude, *The System Builders: The Story of SDC,* Santa Monica, Calif.: System Development Corporation, 1981.

Beard, Edmund, *Developing the ICBM,* New York: Columbia University Press, 1976.

Blanchard, Benjamin S., and Wolter J. Fabrycky, *Systems Engineering and Analysis,* Upper Saddle River, N.J.: Prentice Hall, 1998.

Buede, Dennis M., *The Engineering Design of Systems, Models and Methods,* New York: Wiley, 2000.

Coffin, M. D., and Charles F. Tiffany, "New Air Force Requirements for Structural Safety, Durability, and Life Management," *Journal of Aircraft,* Vol. 13, No. 2, February 1976, pp. 93–98.

Defense Science Board, *Corrosion Control*, Washington, D.C., October 2004.

Defense Systems Management College, *Systems Engineering Fundamentals,* Fort Belvoir, Virginia: Defense Acquisition University Press, January 2001.

Duncan, Francis, *Rickover and the Nuclear Navy, The Discipline of Technology,* Annapolis, Md.: Naval Institute Press, 1990.

Forman, Royce G., "Early History and Current Development Efforts in Fracture Mechanics Applications for Aircraft," Savannah, Ga.: 2002 USAF Aircraft Structural Integrity Program Conference, December 10–12, 2002.

Gebman, Jean R., and Paul C. Paris, *Probability That the Propagation of an Undetected Fatigue Crack Will Not Cause a Structural Failure*, in *Fatigue Crack Growth Measurement and Analysis*, American Society for Testing and Materials, STP-738, 1979.

———, *Probability That the Propagation of an Undetected Fatigue Crack Will Not Cause a Structural Failure*, Santa Monica, Calif.: RAND Corporation, R-2238-RC, June 1978. As of August 1, 2007: http://www.rand.org/pubs/reports/R2238/

Government Electronics and Information Technology Association (GEITA), *Processes for Engineering a System,* ANSI/GEIA EIA-632, September 2003.

Hahn, Gail, Boeing, "Accelerated Insertion of Materials—Composites: A Technology Investment Agreement," presentation at the MMS-OTRC Workshop: Qualifying New Technology for Deepwater Oil and Gas Development, October 29, 2002a.

———, "Accelerated Insertion of Materials—Impact of Manufacturing on Performance," presentation at the MMS-OTRC Workshop: Qualifying New Technology for Deepwater Oil and Gas Development, October 29, 2002b.

Institute of Electrical and Electronics Engineers, *Standard for Application and Management of the System Engineering Process,* IEEE Standard 1220-1998, New York, 1998.

International Council of Systems Engineers, *Systems Engineering Handbook,* Seattle, Washington, July 2000.

Kim, Yool, Stephen Sheehy, and Darryl Lenhardt, *A Survey of Aircraft Structural-Life Management Programs in the U.S. Navy, the Canadian Forces, and the U.S. Air Force,* Santa Monica, Calif.: RAND Corporation, MG-370-AF, 2006. As of August 1, 2007:
http://www.rand.org/pubs/monographs/MG370/

Lincoln, John W., "Aging Aircraft Issues in the United States Air Force," *SAMPE Journal,* Vol. 32, No. 5, 1996, pp. 27–33.

———, *Risk Assessments of Aging Aircraft,* Ogden, Utah, DoD/FAA/NASA Conference on Aging Aircraft, July 8–10, 1997.

Maier, Mark W., Eberhardt Rechtin, *The Art of Systems Architecting,* Washington, D.C.: CRC Press, 2000.

MIL-HDBK—*See* Military Handbook.

MIL-STD—*See* Military Standard.

Military Handbook 17/1F, *Composite Materials Handbook, Vol. 1, Polymer Matrix Composites Guidelines for Characterization of Structural Materials,* 2002.

———, 17/2F, *Composite Materials Handbook, Vol. 2, Polymer Matrix Composites Materials Properties,* 2002.

———, 17/3F, *Composite Materials Handbook, Vol. 3, Polymer Matrix Composites Materials Usage, Design, and Analysis,* 2002.

———, 17/4A, *Composite Materials Handbook, Vol. 4, Metal Matrix Composites,* 2002.

———, 17/5, *Composite Materials Handbook, Vol. 5, Ceramic Matrix Composites,* 2002.

Military Standard 1530C, *Department of Defense Standard Practice: Aircraft Structural Integrity Program (ASIP),* Washington, D.C.: U.S. Air Force, November 1, 2005.

National Aeronautics and Space Administration, *Systems Engineering Handbook,* SP-610S, Washington, D.C., June 1995.

National Research Council, *Aging of U.S. Air Force Aircraft,* National Materials Advisory Board, Washington, D.C.: National Academy Press, NMAB-488-2, 1997.

National Transportation Safety Board, *In-Flight Separation of Vertical Stabilizer, American Airlines Flight 587, Airbus Industrie A300-605R, N14053, Belle Harbor, New York, November 12, 2001,* AAR-04/04, PB2004-910404, Washington, D.C., 2004.

NRC—*See* National Research Council.

Paris, Paul C., *The Fracture Mechanics Approach to Fatigue,* in J. J. Burke, N. L. Reed, and V. Weiss, eds., *Fatigue—An Interdisciplinary Approach,* Syracuse, New York: Syracuse University Press, 1964, pp. 107–132.

Paris, Paul C., M. Gomez, and W. Anderson, "A Rational-Analytic Theory of Fatigue," *The Trend in Engineering,* Seattle, Wash.: University of Washington, Seattle, 1961.

Porter, William A., *Modern Foundation of System Engineering,* New York: Macmillan, 1968.

Rickover, Hyman, Admiral, U.S. Navy, Director, Naval Nuclear Propulsion Program Statement before the House Subcommittee on Energy and Propulsion, May 1979.

Rockwell, Theodore, *The Rickover Effect, How One Man Made a Difference,* Annapolis, Maryland: Naval Institute Press, 1992.

Rouse, William B., and Kenneth R. Boff, "Value-Centered R&D Organizations: Ten Principles for Characterizing, Assessing, and Managing Value," *System Engineering,* Vol. 7, No. 2, Hoboken, New Jersey: Wiley, 2004, pp. 167–185.

Sage, Andrew P., *Systems Engineering: Methodology and Applications,* New York: IEEE Press, 1977.

Sage, Andrew P., and James A. Melsa, *System Identification,* New York: Academic Press, 1971.

Sapolsky, Harvey M., *The Polaris System Development,* Cambridge, Massachusetts: Harvard University Press, 1972.

Schutz, Walter, "A History of Fatigue," *Engineering Fracture Mechanics,* Vol. 54, No, 2, Great Britain: Elsevier, 1996, pp. 263–300.

Schwartz, Peter, *The Art of the Long View,* New York: Doubleday, 1991.

Taormina, Tom, *Virtual Leadership and the ISO 9000 Imperative,* Upper Saddle River, New Jersey: Prentice Hall, 1996.

U.S. Air Force, *KC-X Request for Proposal, Contract Data Requirements List,* January 2007.

U.S. Department of Defense, *Systems Engineering Plan (SEP), Preparation Guide,* Washington, D.C., February 10, 2006.

U.S. Department of Transportation, Federal Aviation Administration, "Conducting Records Reviews and Aircraft Inspections Mandated by the Aging Aircraft Rules," Notice 8300.113, November 25, 2003.